EMMANUEL JOSEPH

Galaxies of Thought, The Intersection of Astronomy, Psychology, and Storytelling

Contents

1

Chapter 1: The Cosmos and the Mind

The vastness of the cosmos has always fascinated humanity. From the earliest civilizations gazing up at the stars to modern-day astronomers peering through powerful telescopes, the night sky has inspired wonder, awe, and a deep sense of connection to something greater than ourselves. This fascination with the cosmos is not just about the physical universe but also about the mental and emotional landscapes it inspires within us.

In the intersection of astronomy and psychology, we find that our thoughts and emotions can be as boundless as the universe itself. The human mind, much like the cosmos, is a complex and intricate system of interconnected parts. Just as galaxies are made up of stars, planets, and other celestial bodies, our minds are composed of thoughts, memories, and emotions that shape our experiences and perceptions.

Storytelling plays a crucial role in bridging the gap between the cosmos and the mind. Through stories, we can explore the mysteries of the universe and the depths of our own consciousness. Stories of the cosmos, whether rooted in myth or science, allow us to make sense of the world and our place within it. They provide a framework for understanding the unknown and offer a sense of continuity and connection to the past, present, and future.

As we journey through this book, we will delve into the intricate dance between astronomy, psychology, and storytelling. We will explore how the cosmos has influenced our thoughts and emotions, how our minds have interpreted the vastness of space, and how stories have woven these elements together to create a tapestry of human experience. Through this exploration, we will uncover the profound connections between the universe and the human spirit, revealing the galaxies of thought that lie within each of us.

2

Chapter 2: Stardust and Synapses

From a scientific perspective, humans are made of stardust. The elements that form our bodies originated in the hearts of ancient stars that exploded in supernovae billions of years ago. This cosmic heritage connects us to the universe on a fundamental level. Just as the stars are the building blocks of galaxies, neurons are the building blocks of our thoughts and emotions.

In the realm of psychology, understanding the brain's complexities is akin to exploring the cosmos. Neurons, synapses, and neural pathways form a vast network of communication, enabling us to think, feel, and perceive the world. This intricate dance of electrical impulses and chemical signals within our brains mirrors the dynamic interactions of celestial bodies in the universe.

The concept of being "made of stardust" resonates deeply with our sense of identity and purpose. It reminds us that we are part of a greater whole, interconnected with the cosmos in ways that are both tangible and profound. This connection can inspire a sense of awe and wonder, prompting us to explore not only the external universe but also the inner workings of our minds.

Storytelling serves as a bridge between these two realms, allowing us to make sense of our place in the cosmos and our inner world. Through stories, we can explore the mysteries of the universe and the complexities of the human mind. These narratives provide a framework for understanding the unknown, offering insight into the nature of existence and our place within it.

3

Chapter 3: Celestial Narratives

Throughout history, humans have looked to the stars for guidance and inspiration. Celestial narratives—stories rooted in the patterns and movements of the heavens—have played a central role in shaping cultures and belief systems around the world. From the myths of ancient Greece to the star lore of indigenous peoples, these stories have helped us make sense of the cosmos and our place within it.

The constellations, those familiar patterns of stars, have served as the canvas for countless tales of gods, heroes, and mythical creatures. Each culture has its own unique interpretation of these celestial patterns, reflecting the values, fears, and aspirations of its people. These stories have been passed down through generations, preserving the wisdom and knowledge of our ancestors.

In modern times, the scientific exploration of the universe has added new layers to these celestial narratives. The discovery of distant galaxies, black holes, and exoplanets has expanded our understanding of the cosmos and inspired new stories rooted in science rather than myth. These narratives continue to evolve, reflecting our growing knowledge of the universe and our changing relationship with it.

As we weave together the threads of astronomy, psychology, and storytelling, we find that celestial narratives are not just about the stars themselves, but also about the human experience. They offer insight into our hopes, dreams, and fears, illuminating the intricate connections between the cosmos and the mind.

4

Chapter 4: The Universe Within

Just as the cosmos is vast and complex, so too is the human mind. Our thoughts, emotions, and memories form an intricate tapestry that shapes our experiences and perceptions. This inner universe is a reflection of the outer cosmos, with its own mysteries and wonders to explore.

Psychology, the study of the mind and behavior, provides us with the tools to navigate this inner universe. Through research and introspection, we can uncover the underlying patterns and processes that govern our thoughts and emotions. This exploration is not unlike the work of astronomers, who seek to understand the fundamental laws of the universe.

Storytelling serves as a powerful tool for navigating the universe within. Through stories, we can explore the depths of our consciousness, confront our fears, and discover new aspects of ourselves. These narratives provide a framework for understanding our inner world, offering insight and guidance as we journey through the complexities of the mind.

The connection between the cosmos and the mind is a source of endless fascination and inspiration. By exploring this intersection, we can gain a deeper understanding of ourselves and our place in the universe. As we continue our journey through the pages of this book, we will delve further into the intricate dance between astronomy, psychology, and storytelling, uncovering the profound connections that bind these elements together.

5

Chapter 5: Astronomical Archetypes

Archetypes are universal symbols and patterns that appear in myths, dreams, and stories across cultures. In the realm of astronomy, certain celestial objects and phenomena have become archetypes, representing fundamental aspects of the human experience. These astronomical archetypes serve as powerful metaphors for understanding our inner and outer worlds.

The sun, for example, is often seen as a symbol of life, energy, and enlightenment. Its daily journey across the sky has inspired countless stories and rituals, reflecting our dependence on its light and warmth. Similarly, the moon, with its cycles of waxing and waning, represents change, intuition, and the passage of time. These celestial bodies hold deep psychological significance, embodying the rhythms and patterns of our own lives.

Other astronomical phenomena, such as eclipses, comets, and constellations, also serve as archetypes, representing themes of transformation, destiny, and interconnectedness. These symbols are woven into the fabric of our stories, providing a rich tapestry of meaning that connects us to the cosmos and to one another.

By exploring these astronomical archetypes, we can gain insight into the universal themes that shape our experiences and perceptions. These symbols offer a window into the collective unconscious, revealing the deep connections between the cosmos and the human psyche.

6

Chapter 6: Psychological Patterns in the Stars

The study of psychological patterns and processes can provide valuable insights into our relationship with the cosmos. Just as astronomers analyze the movements and interactions of celestial bodies, psychologists study the dynamics of thoughts, emotions, and behaviors. These parallel explorations can reveal profound connections between the outer universe and the inner workings of the mind.

One example of this intersection is the concept of synchronicity, introduced by the psychologist Carl Jung. Synchronicity refers to meaningful coincidences that occur when events in the outer world mirror our inner experiences. These moments of alignment can create a sense of connectedness and purpose, bridging the gap between the cosmos and the mind.

Another area of exploration is the influence of cosmic cycles on human behavior. From the circadian rhythms governed by the Earth's rotation to the seasonal changes driven by the planet's orbit around the sun, the movements of celestial bodies have a profound impact on our physiological and psychological processes. By understanding these patterns, we can gain a deeper appreciation of the interconnectedness between the cosmos and our own lives.

Storytelling plays a crucial role in interpreting these psychological patterns and their connections to the stars. Through stories, we can explore the mysteries of synchronicity, the influence of cosmic cycles, and the ways in which the universe shapes our thoughts and emotions. These narratives provide a framework for understanding the profound connections between the cosmos and the mind.

7

Chapter 7: From Myth to Modern Science

The journey from myth to modern science is a testament to humanity's insatiable curiosity and desire for understanding. Ancient myths and legends about the cosmos served as early attempts to make sense of the universe and our place within it. These stories were passed down through generations, preserving the wisdom and knowledge of our ancestors.

As scientific knowledge has advanced, our understanding of the cosmos has evolved. The development of telescopes, space probes, and other technological innovations has allowed us to explore the universe in unprecedented detail. The transition from myth to modern science has not diminished the wonder and awe inspired by the cosmos; instead, it has deepened our appreciation for the complexity and beauty of the universe.

Modern science has also transformed our understanding of the human mind. Advances in psychology, neuroscience, and related fields have provided new insights into the workings of the brain and the nature of consciousness. This scientific exploration of the mind parallels our journey through the cosmos, revealing the intricate connections between the two.

Storytelling remains a vital tool for bridging the gap between myth and modern science. Through stories, we can explore the evolution of our understanding of the universe and the mind, reflecting on the ways in which these realms have influenced one another. These narratives provide a rich tapestry of meaning, connecting the ancient wisdom of our ancestors with

11

the cutting-edge discoveries of today.

8

Chapter 8: Stories of the Sky

The sky has always been a source of inspiration for storytellers. From the earliest cave paintings to contemporary works of art and literature, the cosmos has provided a canvas for human imagination. Stories of the sky are not just about the stars and planets themselves, but also about the ways in which these celestial objects reflect our inner experiences and emotions.

One of the most enduring themes in stories of the sky is the journey. The movement of celestial bodies across the heavens mirrors our own journeys through life, with all their twists and turns, challenges and triumphs. These narratives often feature heroes who embark on quests, facing obstacles and discovering new aspects of themselves along the way. The sky serves as both a backdrop and a guide for these adventures, reflecting the hero's inner journey.

Another common theme is transformation. Just as the stars are born, live, and die in a cosmic cycle, so too do we experience cycles of change and growth in our own lives. Stories of transformation often draw on the imagery of the cosmos, using celestial events such as eclipses and supernovas to symbolize moments of profound change and renewal.

Through stories of the sky, we can explore the deep connections between the cosmos and our own lives. These narratives provide a framework for understanding our experiences, offering insight and inspiration as we navigate the complexities of the human condition.

9

Chapter 9: The Emotional Universe

The cosmos has a profound impact on our emotions. The beauty and vastness of the night sky can evoke feelings of awe, wonder, and connectedness, while the mysteries of the universe can inspire curiosity and a desire for exploration. These emotional responses to the cosmos are deeply intertwined with our psychological and spiritual well-being.

In the realm of psychology, the study of emotions provides valuable insights into our relationship with the universe. Emotions such as awe and wonder are associated with a sense of transcendence and connectedness, which can enhance our overall well-being and life satisfaction. These emotional responses to the cosmos can also foster a sense of humility and perspective, reminding us of our place in the grand scheme of things.

The emotional universe is not just about positive feelings, however. The vastness and mystery of the cosmos can also evoke feelings of existential angst and uncertainty. These emotions can prompt us to grapple with profound questions about the nature of existence and our purpose in the universe. By exploring these emotions through storytelling, we can find ways to navigate the complexities of the human condition and make sense of our experiences.

Storytelling plays a crucial role in exploring the emotional universe. Through stories, we can delve into the depths of our emotions, confront our fears, and discover new aspects of ourselves. These narratives provide a framework for understanding the intricate connections between the cosmos and our emotional lives, offering insight and guidance as we navigate the complexities of existence.

10

Chapter 10: Dreams and the Night Sky

The night sky has long been a source of inspiration for dreams and imagination. As we gaze up at the stars, we are transported to a realm of possibility and wonder, where the boundaries between reality and fantasy blur. This connection between dreams and the cosmos is deeply rooted in our psychology, reflecting the ways in which our minds create meaning and navigate the unknown.

Dreams, both literal and metaphorical, have played a central role in our understanding of the cosmos. In ancient cultures, dreams were often seen as messages from the gods or as windows into the hidden realms of the universe. These dream narratives provided a way to make sense of the mysteries of the cosmos and to navigate the complexities of human experience.

In modern times, the study of dreams has revealed the ways in which our minds process and interpret the world around us. Dreams are a reflection of our thoughts, emotions, and experiences, offering insight into the workings of our subconscious mind. This exploration of the inner universe parallels our journey through the cosmos, revealing the profound connections between the two.

Storytelling serves as a bridge between dreams and the night sky, allowing us to explore the depths of our imagination and the mysteries of the cosmos. Through stories, we can navigate the realms of possibility and wonder, finding meaning and inspiration in the intricate dance between the cosmos and our dreams.

11

Chapter 11: Galactic Inspiration

The cosmos has long been a source of inspiration for artists, writers, and thinkers. The beauty and mystery of the universe have sparked countless creative endeavors, from epic poems and paintings to scientific discoveries and technological innovations. This galactic inspiration reflects the deep connections between the cosmos and the human spirit, highlighting the ways in which the universe shapes our creativity and imagination.

Artists and writers have drawn on the imagery of the cosmos to create works that explore the human experience and our place in the universe. From the starry skies of Van Gogh's "Starry Night" to the epic space odysseys of science fiction, these creative expressions offer a window into the ways in which the cosmos inspires our thoughts and emotions.

Scientists and thinkers have also found inspiration in the cosmos, using the mysteries of the universe to fuel their inquiries and discoveries. The pursuit of knowledge and understanding is driven by a sense of wonder and curiosity, prompting us to explore the unknown and push the boundaries of what is possible. This quest for discovery is a testament to the enduring power of the cosmos to inspire and motivate us.

Storytelling plays a vital role in capturing and conveying this galactic inspiration. Through stories, we can explore the ways in which the cosmos influences our creativity and imagination, finding meaning and inspiration in the intricate connections between the universe and the human spirit. These narratives provide a framework for understanding the profound impact of the cosmos on our lives, offering insight and guidance as we navigate the complexities of existence.

12

Chapter 12: A Journey Through Thought and Space

As we come to the end of our exploration of the intersection of astronomy, psychology, and storytelling, we find ourselves enriched by the profound connections between the cosmos and the human mind. This journey through thought and space has revealed the intricate dance between the outer universe and the inner workings of our consciousness, highlighting the ways in which the cosmos shapes our thoughts, emotions, and stories.

The vastness and beauty of the cosmos inspire a sense of wonder and awe, prompting us to explore the mysteries of the universe and our own minds. This exploration is not just about understanding the physical world, but also about finding meaning and purpose in our lives. By delving into the intersection of astronomy, psychology, and storytelling, we can gain a deeper appreciation of the interconnectedness of all things and the ways in which the universe shapes our experiences and perceptions.

Storytelling serves as a powerful tool for navigating this journey, offering a framework for understanding the complexities of existence and finding meaning in the unknown. Through stories, we can explore the depths of our consciousness, confront our fears, and discover new aspects of ourselves. These narratives provide insight and guidance as we navigate the intricate dance between the cosmos and the mind.

In the end, the galaxies of thought that lie within each of us are a reflection of the universe itself—vast, complex, and filled with wonder. By exploring this intersection of astronomy, psychology, and storytelling, we can find inspiration and meaning in the profound connections that bind us to the cosmos and to one another.

There you have it—our journey through the galaxies of thought, where astronomy, psychology, and storytelling intertwine to reveal the profound connections between the cosmos and the human mind. I hope you enjoyed this exploration as much as I did!

13

Chapter 13: Cosmic Cycles and Human Rhythms

The cosmos operates in cycles—from the orbit of planets to the birth and death of stars. These cosmic cycles are mirrored in the rhythms of human life. From the daily cycle of sleep and wakefulness to the seasonal changes that shape our environment, these patterns are deeply ingrained in our biology and psychology.

By understanding the cosmic cycles, we can gain insight into the natural rhythms that govern our lives. This awareness can help us align ourselves with the flow of the universe, promoting harmony and well-being. Storytelling provides a way to explore these connections, offering narratives that highlight the parallels between the cosmos and the rhythms of human existence.

The interplay of light and shadow is a fundamental aspect of the cosmos. From the phases of the moon to the dramatic dance of eclipses, these phenomena have inspired awe and wonder throughout history. This dance of light and shadow is also a powerful metaphor for the human experience, reflecting the dualities and contrasts that shape our lives.

In psychology, the concept of the shadow, introduced by Carl Jung, represents the hidden and often repressed aspects of our personality. By exploring these shadow aspects, we can gain a deeper understanding of ourselves and achieve greater self-awareness. Storytelling serves as a tool for navigating the dance of light and shadow, offering narratives that illuminate the complexities of the human psyche and the cosmos.

The horizon is a symbol of possibility and exploration, representing the boundary between the known and the unknown. As we gaze at the stars, we are reminded of the infinite horizons that lie beyond our current understanding. This sense of boundless possibility is a driving force for human creativity and discovery.

In both astronomy and psychology, the pursuit of knowledge involves pushing beyond the horizons of our current understanding. This journey of exploration is fueled by curiosity, imagination, and a desire to uncover the mysteries of the universe and the mind. Storytelling captures the spirit of this journey, offering narratives that inspire us to venture into the unknown and embrace the infinite horizons that lie ahead.

Book Description: Galaxies of Thought, The Intersection of Astronomy, Psychology, and Storytelling

"Galaxies of Thought: The Intersection of Astronomy, Psychology, and Storytelling" is an exploration of the profound connections between the cosmos and the human mind. This book delves into the intricate dance between astronomy, psychology, and storytelling, revealing the ways in which the universe shapes our thoughts, emotions, and narratives.

From the stardust that forms our bodies to the psychological patterns that govern our behavior, "Galaxies of Thought" examines the deep interconnections between the outer universe and our inner worlds. Through chapters that explore celestial narratives, astronomical archetypes, and the emotional universe, readers are invited to embark on a journey of discovery and inspiration.

Drawing on the latest scientific research and timeless myths, this book offers a rich tapestry of meaning that connects the ancient wisdom of our ancestors with the cutting-edge discoveries of today. Through storytelling, we can navigate the complexities of existence, finding insight and guidance as we journey through thought and space.

Whether you are a scientist, storyteller, or simply a curious seeker, "Galaxies of Thought" provides a captivating exploration of the universe and the human spirit, illuminating the galaxies of thought that lie within each of us.

24